CHALLENGER REMEMBERED

HOMETOWN PRESS, INC.
HUNTSVILLE, ALABAMA

Hometown Press, Inc.
2007 Gallatin Street • Huntsville, Alabama 35801
205-539-3320

For the families of Dick Scobee, Mike Smith, Judy Resnik, Ron McNair, Ellison Onizuka, Greg Jarvis, and Christa McAuliffe

This book is a collection of art, poetry and prose from young people who were touched by the *Challenger* disaster. We hope the selections published here are an accurate reflection of the national mood in the aftermath of the tragedy.

Dottie Main
Jeffrey C. Hindman, M.D.
May, 1986

Portraits by
Tom Fricker
Alabama Space and Rocket Center
Huntsville, Alabama

"They had a hunger to explore the universe and discover its truths. They wished to serve, and they did. They served all of us. The Challenger crew was pulling us into the future...
Nothing ends here. Our hopes and our journeys continue."

Ronald Reagan
January 28, 1986

I

Future Astronauts

The stars are my light . . .

Astronaut

I have done what no others
do in a lifetime

I broke through the clouds
and entered the world of
celestial bodies

The stars are my light
the planets are my friends,
their moons are their children

I see no darkness,
only the light of exploration

The father of it all is the sun.

Russ Stanners
Grade 6
Academy for Academics & Arts
Huntsville, Alabama

1/28/86

Dear Mr. President,

My name is Tom Malone. I am a sixteen year-old Junior at the Galloway School in Atlanta, Georgia. I am writing to you because of the unbelievably tragic accident of Space Shuttle Challenger this morning.

This morning, when the news of the disaster was announced at my high school, almost everybody crowded around the television in the teacher's lounge and watched in horror at the replays of the tragic launch. Along with many of my friends, I cried, and cried several more times throughout the day as I continued to watch the newscasts. I mourn for the death of the seven brave human beings who were the crew of this, the twenty-fifth Space Shuttle launch, and I grieve with their families and loved ones. The day January 28th, 1986 will not long be forgotten.

There is a lesson to be taken from this tragedy; indeed from any tragedy, and that is that human beings are not invincible, nor are they immortal. It is a reminder that life is not always perfect, and we do not live in Utopia. But people who never take chances never receive chances, and human history has shown that the greatest gain, in any field, not just science, comes when humans take that chance, become brave in the face of danger and adversity, and indeed risk their very lives for gain; gain they consider so important that they could risk their lives for it. If somehow the seven Challenger astronauts could be asked not if they would fly in a space shuttle again, they would certainly answer with a definite "yes".

I want you to know that although students are sad at the loss of human life, they have not lost faith in the American Space Program, and wish it to continue stronger than ever. I am a member of a group of six students from all over the country participating in a NASA student involvement project called Project Explorer, under the direction of Dr. Konrad K. Dannenberg of Marshall Space Flight Center and the Alabama Space and Rocket Center. I am developing a scientific experiment that will be flown aboard a space shuttle. This experiment corresponds with my deepest interest in science and the space program; this is where I hope to make my career in the coming years of my life. I think it is important for the people of the United States, and indeed the world, to know that the young people of America are enthusiastic about the American Space Program, and are interested and involved.

It is my belief that the future of humanity will be irrevocably connected and sustained by space and human utilization of it, both for industry and research, but that more importantly, humanity will be brought into the future by the courage and love of challenge that is so importantly and symbolically shown by the Shuttle Challenger and its crew. This goal can only be reached by such pioneers as the Challenger crew who died in today's tragic accident. We should not lose faith or hope because of a setback, grievous though it is. Indeed it should serve as an incentive to strive harder and further in our efforts on behalf of humankind. I think the next launch of the Space Shuttle should be on schedule, as planned. I believe that indeed this is what the astronauts aboard Challenger would have wanted.

Although I am a student, I think it important that I say seven astronauts, even though one of them was a teacher. To me, they were a crew of seven human beings, and that regardless of their individual profession, they all had one underlying commitment in mind; to further the education and condition of mankind. I believe that this is what the astronauts would have wanted: to continue on in their footsteps in the quest for knowledge. This, after all, is what life is really all about.

Thank you very much for your time.

Sincerely,

Thomas Patrick Malone II (Tom)
Grade 11
Galloway School
Atlanta, Georgia

3/10/86

Gentlemen:

I am really sad about the Challenger and the crew. I am interested in some posters about space. My brother has a model and a poster of a space shuttle. I wish I could be the second kid in space. Our class studied about the planets. My favorite planet is Saturn. I did a report on Mercury.

Well, good-bye.

Sincerely,

Carrie Ann Miller
Grade 4
Greencastle-Antrim
Greencastle, Pennsylvania

1/28/86

My friend Edward Buckbee,

Today I woke up and started to watch the launching of mission 51-L. The terrible tragedy that I watched unfold started my thinking about my dreams to be an astronaut. I will become an astronaut someday no matter what. Something like what happened today is "Part of the Job". It will not scare or discourage me.

I am writing this letter at the spur of the moment (so please excuse the printing and errors) to tell you that we Americans must never stop! in exploring space.The Space Camp program is great! Keep up the good work! So as we forge ahead in the manned space program let us think of the words of Dr. Robert Goddard, a space pioneer like you, and like I hope to be or should I say with great desire will be, "There can be no thought of finishing, for aiming at the stars, both literally and figuratively, it's a problem to occupy generations, (mine and yours!)."

So that, no matter how much progress one makes, there is always the thrill of just beginning. You have begun and I shall continue. America is the best and we must never stop! Go for it!

The Dream is Alive!
Forever!

Winston Kettle
Space Camper and future astronaut.
I hope!
Grade 10
Jefferson High School
Bloomington, Minnesota

To Whom it May Concern:

I am 16 years old and a junior at Arab High School. I am very interested in the Space Program but was very sorry to hear about the tragic mishap that happened Tuesday, January 28, 1986. I have followed the Space Program since the first launching of Columbia. I plan to enter this particular field after I graduate. I don't think we should discontinue the program we have started.

I would appreciate it if you would send me stickers of all the shuttles that have launched, especially Tuesday's launching. I would love to have a picture of the Challenger crew that lost their lives to further our exploration and discovery. Judith Resnik was one lady who I really looked up to. I wanted to be just like her, and still do. If there is any way, whatsoever, you could send me what I've requested, I will be forever grateful. I'm willing to pay any price for this precious material. Please answer my letter.

Thank you for your time,

Sonia Norris
Grade 11
Arab High School
Arab, Alabama

2/26/86

Dear NASA,

I am Jennifer Brockway from New Hope School. We are studying about space in my 5th grade science period. Would you please send me some booklets, pictures, packets, anything you have on space or anything that has to do with space. I would like to be an astronaut when I grow up.

I stress my grief over the Space Shuttle exploding. But I would go up in a Heart Beat!

It's true the space program must go on. Perhaps one day I will fulfill my goal to be an astronaut and work for NASA. Then I will get all I've ever wanted, to explore a whole new world.

Sincerely,

Jennifer Brockway
Grade 5
New Hope, Alabama

P. S. Would you also please tell me my best bet to be an astronaut.

February 15, 1986

Dear NASA,

My name is Jack R. Franks. I am a student at Legion Park School in Houma, Louisiana. I was watching the launch of the space shuttle Challenger on January 28, 1986. It was the most horrible moment I ever saw in space history. When all 7 crew members were killed in that horrible explosion.

I was talking to my school teacher, my best, Mrs. Valvaro, and I asked, "Do you think that they will send the 'first boy in space'?" and she said, "I bet that they will send one up someday." I will like to be the first on the new Challenger. If I do be the first, I am doing it for the seven crew members who lost their lives on that horrible explosion. Especially the teacher Christa McAuliffe, and every child on this earth.

This mission was especially for children. In which all children would know what space is like. I was concerned about this special mission. I couldn't wait until Mrs. McAuliffe told her lessons on the shuttle.

Last year I was selected to go to Space Camp in Huntsville. And I need to ask a favor. I would like to know if you can schedule an appointment for me to go talk to an official at the Marshall Space Flight Center? I have a way to get there but I would like to go talk to you about my feelings of the new Challenger and why I would like to be the first boy in space. Would you write back please. And I would like to know if you could send me any posters or pictures of the space shuttles and their crew members. And please send me a picture of Mrs. Christa McAuliffe, and all of the crew members of the Challenger who got killed in that explosion.

Keep up the good work and I say my prayers for you.

God Bless You.

Jack Franks
Grade 7
Legion Park School
Houma, Louisiana

2/14/86

Dear NASA,

I am so sorry about what happened to the Space Shuttle Challenger. I am so interested in the space shuttle that I would go into space tomorrow no matter what happened. If you would, I would really like to learn more about the shuttle. Could you please send me some information about the shuttle.

A Space Program Lover,

John Lassiter
Grade 4
Mobile, Alabama

P.S. I am in the Young Astronaut program too.

Dear Company,

I permit myself to write you as one correspondent among thousands who are following NASA's activities since this exists.
Would you apologize to disturb you in that fatal period of the 86's program but I would just say one word to show you how we all understood what just happened today with the shuttle Challenger II.
Remember what said Neil Armstrong from the moon on 21st July 1969 at 3h56: "One step for man, one giant leap for mankind."
So nothing is lost and shuttle will be launch again and again because that's a need for humanity.
Many thanks to have read this letter.

Sincerely,

Steve Balas
Annecy
France

ATTN: Director, Space Camp

Dear Director,

In August, 1983, my son, Keith M. Warman, attended Space Camp. It was a fantastic and enjoyable experience for him. I remember on his last day of camp, we were told there was an interest in following the attendees as they mature.

In January of this year, Keith was required to write several poems for a literature class project. Enclosed is a copy of one of his poems. It was written in mid-January before the Challenger disaster.

Even though Keith has become involved in several other activities in high school, he is still interested in the space program and tries to keep up with what is happening. When he showed me the poem, I was sure you should know about it also.

Please accept my appreciation for your kind attention.

Sincerely,

Mrs. Christine Warman

"Spacewalk" is a poem based on a true story of Astronaut Edward White II which occurred in June of 1965. On the Gemini 4 spacecraft Ed White became America's first astronaut to leave the spacecraft. Both NASA and the other pilot had a lot of problems trying to get him to come back in.

Spacewalk

Out to Space, the last frontier.
It's time to open the door,
and let the starlight appear.
In his pressured suit, gold visor over his eyes,
He floats out of the space craft,
looking dazzled and surprised.

Suspended in space, he looks down on the world
The oceans so blue
and the clouds so swirled.
"Let's go Ed, come on back
We've got a million miles to go,
and a million miles on our back!"

He can't do it, he won't go back in
It's so wonderful out there,
Only he knows how its been.
On his way in, he takes a last look at space.
Then the door shuts,
and a tear runs down his face.

Keith Warman
Forest, Virginia

To: The Space Rocket Center

I'm writing to tell you that I want to be a good astronaut. I'm really anxious. I saw a boy on T.V., he is on commercials, the one who plays Messy Marvin. He said something about you on T.V. and about the tragedy of the Challenger. I got your address from the World Almanac for Weekly Readers Books - it's called the Kid's World Almanac of Records and Facts.

Well, I want to be an astronaut because I feel like I was meant to be one. I watched the tragedy of the space shuttle but if I ride in a shuttle I won't be afraid. You might think I'm saying this to make you think I'm big. But I really mean it, I'm not afraid. When I found out about the tragedy I cried with the rest of the world. I feel like I should take their place - the ones that died. I pray that there are no further accidents like this in the future.

No matter what stands in my way, I'm going to be an astronaut. Please send me more information for me to join the classes. I would like to join your brave crew.

My name is Shannon McCollam, I am 14 years old. I live in South Carolina. I know I might be too old for it but please let me join it, it's my dream. There is no other way to learn about astronauts.

Thank you,

Shannon McCollam
Grade 8
Socastee Middle School
Dunes, South Carolina

Dear NASA,

My name is Tamara Seraj and I am ten years old. I am so sorry about the shuttle "Challenger". But I believe in the importance of space and the space program. I would like to inquire about sending the first young person into space. If you have been thinking about it and you would like to send one of us in a shuttle, I would like to go. Even after that big tragedy that happened today I would want to go. I have had an interest in flying in a shuttle for a long time.

Sincerely,

Tamara Seraj
Grade 5
Jasper Public School
Rancho Cucamonga, California

Dear Sir,

My name is Mark Vinsant. I am twelve years old. I have been interested in space for a long time.

And I saw the tragic moment when the space shuttle "Challenger" blew up. That was a sad moment.

Well, I heard Mr. Reagan say that there will be flights with volunteers, teachers, and more people. And I was talking with my parents and if you ever need a child to go into space I will be glad to go.

But I also wanted to pray for the seven astronauts who wanted to go up in space and were killed.

And I know the risk and I'm not scared of dying for my country.

Your friend,

Mark Vinsant
Grade 6
Cahaba Heights Comm. School
Birmingham, Alabama

2/29/86

Dear NASA,

We feel bad about the Space Shuttle Tragedy. We would like to fulfill the dreams of Christa McAuliffe, and be a part of her love for outer space. We have always loved space. My friend and I are only twelve years old. We are willing to take any test that you give us. We would very much appreciate it if you would give it some thought about sending two kids into space. If you decide to send two kids into space we would consider it a privilege if you would consider us. Our names are Lynn Gardenhire and Anthony Miller. We live in Russellville, Arkansas 72801. We have collected every bit of information about the space shuttle accident.

Sincerely Yours,

Lynn Gardenhire
Anthony Miller
Grade 12
Russellville, Arkansas

P. S. Anthony wants to know why there is a big spike on the front of the Space Shuttle Enterprise?

Dear Sirs:

Enclosed you will find a poem that was written by one of our fifth grade students after learning of the tragic accident of our Space Shuttle.

Sincerely,

Ruth F. Adams
Principal
Christian Central Academy
Buffalo, New York

A *mericans*
S *triving to explore*
T *rying to discover*
R *emaining untold secrets*
O *nward to the sky*
N *ow we have seven heroes*
A *ll very brave*
U *ntold secrets still lie ahead*
T *ill we find a way*

Steven Sharpe
Grade 5
Christian Central Academy
Buffalo, New York

1/30/86

Dear Sir:

Hi. My name is Alicia. I am fifteen years old and in the ninth grade. I would like to offer my condolences to you on the loss of your space shuttle and the lives of seven good people. This tragedy has helped me decide what I want to do with my life. It does not deter me in the least. I would very much like to become an astronaut. I am especially interested in computers, mainly computer programming and electronics. If you have any advice on what courses I should take in high school and college I would appreciate it. Or anything that would help me become an astronaut.

I wrote to you in Huntsville because you are located in the same state as I and would know of the schools in this area.

Thank you for taking the time to read my letter. I appreciate it very much.

Yours truly,

Alicia Blankenship
Grade 9
Alexander City Jr. High
Alexander City, Alabama

1/31/86

Dear NASA,

I am sorry to hear that the space shuttle blew up. Just think you have got to build another one. Could you send me some information about space.

When I grow up I might be an astronaut. After I saw that I don't know. Was you carrying a telescope.

Your friend,

John Daniel Buttz
Grade 3
Lucas Elementary School
Chariton, Iowa

2/12/86

Dear Marshall Space Flight Center

I am writing in regards to the Space Shuttle. I really was stunned for awhile. I am very sorry that it happened like that.
Would you send me more information on how I can go up in space. I sent letters out to all the Space centers, to the President and all over the U.S. to try to get up in space. I hope I get to go.
Write back soon.

Bob Bryant
Grade 8
Atchinson, Kansas

2/4/86

Dear NASA,

I think you are the best space center in the world. I am sorry that Challenger blew up. I am interested in the space program. I am going to be an astronaut when I grow up. So is my friend Martin Smith.

Stephen Parrow

Grade 2
Hiller Elementary School
Madison Heights, Michigan

3/12/86

Dear Sir or Madam:

Last week when we were watching the T.V. show "Punky Brewster" we saw the segment about the "Young Astronauts" organization. We, along with other kids in the United States, were hurt to think that the explosion of the Space Shuttle Challenger would set us back in space technology. We feel that the space program will, in the future, be handed down to our generation in that we will be carrying on the ideas and thoughts of this program.

We, as individuals, would like to help keep this program going. Could you send us information about "Young Astronauts" and how we could join. Thank you very much.

Sincerely,

Kim Pearsall (13)
Jennifer Graeser (13)
Sidney, Iowa

II

The Spirit of Challenger

Cause Earth can't keep us
here forever . . .

Dear Sir or Madam,

Hi my name is Denise Carter. I am the daughter of Wade Batts. I go to Athens Middle School and am in the 6th grade. I wrote a poem about the Challenger. I really would like for it to be in the Space Center. I know you will get a lot of poems, but this one is very special. I'm sending a letter and poem to the seven astronaut's families. I really would like for the poem to be put in the Space Center. Please!? Also, would you write back. Thank you for your time to read this letter.

Sincerely,

Denise Carter

One of my friends said it was good but she also said it wouldn't be good enough. I think so. I also have a bunch of teachers, friends, relatives, and my father thinks it is great. Thank you again!

The Challenger went up -
The Dream came alive,
The crowds were filled
with joy and pride.
When in a moment of
happiness,
We saw the explosion
of the Dream itself.
As though all we could
do is watch and cry,
But until that day
everyone knew the Dream
was alive.
And still till now we
have to learn, pray, and
remember the Challenger
that day.
But that can't keep us
down,
For one day we'll have
to go,
'Cause earth can't keep
us here forever.

Denise Carter
Grade 6
Athens Middle School
Athens, Alabama

2/17/86

Dear Sirs:

My deepest sympathy for the loss of our seven brave astronauts. I hope that our future achievements in space will stand as monuments to the spirit of the Challenger 7 and that this spirit becomes the spirit of all our people, of our species.

I was touched by this tragedy as all Americans were. I'm only a worker in a factory but it made me feel proud the day after the tragedy in the lunchroom to hear my co-workers say, "Let's find out what went wrong, take care of it, build another one (shuttle) and get going again." That's the spirit of the Challenger 7.

Sincerely,

Herman Hayden
Garden City, Michigan

1/30/86

Dear Sir:

We are all terribly saddened by the loss of our seven brave astronauts, but even in our grief we realize that the mission for which these fine people gave their lives must go on....

Even as we saw the terrible tragedy we realized we must not turn back, we must not be afraid. Any one of us is willing to go the next time.

With the Challenger went one fourth of our space delivery capability. We cannot afford that loss. So we are going to do what we can to rebuild the Challenger.

Let this shuttle belong to the people and never be used for military payloads. What more fitting memorial could we ever build than a new spacecraft.

We might call our new spacecraft PHOENIX SEVEN for the mythological bird that rose from its own ashes and for the astronauts.

Very truly yours,

Sean W. Green (Age 11)
Eric R. Green (Age 9)
Tricia D. Green (Age 6)
Woodview Elementary School
Grafton, Wisconsin

2/5/86

Ladies and Gentlemen,

I am twenty-one years old and I am studying information at the Berlin Technical University.

I was very shocked when I saw the last liftoff of Challenger, but I think that mankind needs space flights. So in all my sorrow, I have to say: please go ahead, we need your activity.

I hope you could understand my bad English and thanks a lot for the work you will have.

Your's,

Thomas Posadzy
Berlin, West Germany

She Was A Dreamer

She was a dreamer,
This is true
She tried,
Why don't you
You can give, can't you?
Just to make her dream come true

Her dream was to be a success,
Nothing more, nothing less
The thrill of being in that ship,
Getting ready for that wonderful trip
Knowing you were one of the seven,
Flying through the wonderful heaven

This is a dream, dreamed by many,
It cost hard work, but that's plenty,
For this wonderful dream, dreamed by many.

James Bezotte
Grade 6
Mt. Gap Middle School
Huntsville, Alabama

2/7/86

Firenze, Feb. 7, 1986

Ten days have passed from Challenger's tragedy, our tragedy...
I would like to express my consternation for the dreadful death of the crew: Francis "Dick" Scobee, Michael Smith, Judith Resnik, Ellison-Onizuka, Arnold McNair, Gregory Jarvis and Christa McAuliffe.
The world seems to have already forgotten those names - other problems, other thoughts - but our hearts haven't.
People want to stop manned space programs, but they forget that the development of space environment is not possible without men working up there.
So manned flight must keep on: otherwise the crew of the Challenger would have died vainly.

We all have trust in you,

Paolo Fani
Firenze, Italy

2/18/86

Dear Sirs:

I am very interested in the space program. I have a dream of someday working on or in the space shuttle.

In my opinion (and I guess everybody else's) the shuttle explosion was a terrible accident that no one can do anything about. I know that we lost the lives of seven fellow Americans, but I'm sure that they would want us to continue to try and reach as high as we can. In a way, the explosion could have helped. For one thing, from now on when the shuttles go up, everyone will be watching. And yet in another way, the engineers who build the shuttles will try to make bigger and better things to keep another tragedy from happening. I don't really know how to put into words the way I feel.

Sincerely,

Dalana D. Haynes
Grade 8
Lupton Jr. High
Carbon Hill, Alabama

3/2/86

Dear Sir,

I'm very sorry that the Challenger blew up last Tuesday. Also about Mrs. Christa McAuliffe and her family and her class at school in Florida. At England here in our classroom we are doing a project about space and we are making paper planes called Columbia but we are not going to stop our project because we are sad, we are going to carry on and on and learn more about space!

Yours Sincerely,

Tom Danvers
Jack Bell
Grade 9
Callington, Cornwall
England

Challenger

"We have a lift off" and the people cheer
Not knowing we were to shed a tear
The wives and the parents they stand so proud
Even the children are in the crowd
Now the Challenger has cleared the pad
A time to be happy soon to be sad
The shuttle rises her rockets rumble
And then the flash that turns us humble
We look back and study a flame
Looking for someone or something to blame
We'll solve the problem, and have a delay
There will be a lift off again, they would have wanted it that way.

Sandra Esler
Grade 6
Madison Middle School
Madison, Alabama

I Wonder

When I heard about the explosion of the 7 astronauts, I thought to myself in shock, Why did it happen, why this time? I didn't know much about it. I didn't know about the social studies lessons the teacher was going to teach. After that I thought how exciting it would of been.

My mom said she wanted to go up in space and explore it before the explosion, but after the explosion, she still wanted to go up. She said that just because there are car wrecks we still drive in cars, and the same with airplanes and trains.

If my mom does ever go up then I wonder if she will be scared, she sure is when she rides on a rollercoaster.

Fonda Kirkland
Grade 4
Roger Chaffee Elementary School
Huntsville, Alabama

2/6/86

Sirs,

I am sending you $2.00 so you can have more money to go on with the space work. I will try to get more money at school to help you out.

Lawrence Kelley
Grade 7
Robinson Middle School
Aurora, Missouri

2/10/86

Dear Sirs,

I got $3.00 from my class. I might get more money so here is the money I got today. I hope you get enough money to build the space ship back up to the normal size that it was.

Yours,

Lawrence Kelley
Aurora, Missouri

Challenger

Space "Challenges" each person on Earth
to break its boundaries,
And each person accepts that "Challenge."

It takes courage to break these boundaries
and to go somewhere in something you've
never been.

It also takes adventure in your blood to risk
your life on a journey to man's future.

Jenny Bennett
Grade 7
Mt. Gap Middle School
Huntsville, Alabama

A Debt We Owe

When I heard about the Challenger's terrible destruction, I was stunned. One of my friends was saying it had exploded. I didn't understand. So I smiled. Later the teacher explained what had happened, but it took a while to sink in. The space program is a big part in our lives because we might not live where we do, or have the medicine or technology we use if we didn't have it.

As I think about it I think what Challenger means; to challenge! And the Challenger's crew was challenged in every way one could imagine! As Amelia Earhart once said, "I want to do it because I want to do it. Women must try to do things men have tried. When they fail, then failure must bring but a challenge to others."

Emily Lynch
Grade 6
Mt. Gap Middle School
Huntsville, Alabama

Reaching For A Star

Reaching for a star
A dream to travel fast
and far.
Moving at approximately
"God speed,"
To fulfil a secret need.
Reaching for a star.

What brave and courageous
souls they are
To want to find that
distant star.
A journey made to worlds
on high,
Soaring so far through
the sky.
Reaching for a star.

Kim Brantley
Grade 8
Whitesburg Middle School
Huntsville, Alabama

The Challenge

The problem is not always what it seems
The problem is our fears instead of our dreams.
We've had all our problems, we've had our setbacks.
We've had our mistakes and our miscalculated facts.

But we must continue to travel farther in space,
And now more than ever to quicken the pace.
For I know in my heart that these seven would
want it that way:
To be out in space and be there to stay.

Joe Harchanko
Grade 9
Virgil Grissom High School
Huntsville, Alabama

III

Reactions

The day started out bad . . .

The Day Started Out Bad

I

The day started out bad -
I overslept,
and woke up tired.
A shower didn't help.
The cat had used
a teabag as a ball
But it didn't survive -
Bursting its contents
all over the kitchen floor.

II

Everything started out fine.
In the pilot's voice was calm,
Nothing to indicate
what was to be:
A small explosion
Who knows where
The shuttle
(like the teabag)
Burst its contents
all over the sky
White trails following
The falling debris.
"Obviously a major malfunction."

III

There it was
again and again on TV -
The disbelief in the faces,
The cries of realization,
Tears and screams
of those watching.
All afternoon there was
no escape.
I did the dishes, trying
to forget,
wanting to
put everything down on paper;
the "accident,"
the confusion,
the shock,
But there were no words
inside of me.
People watching prayed
"Oh my God, no!"
What else was there to say?

IV

The day after
and still no escape.
Reporters have digested the news,
Headlines belch the results.
Flags are at half mast.
In LA the great Olympic torch is lit
in response to thousands of calls
asking for something to be done.
People with dazed expressions
and throttled voices
express the nation.
Something tragic has happened.

Karen Pugsley
College Student
Lincoln, Nebraska

Tragedy In Space

Piercing the blue sky on Tuesday
A ship with precious cargo
Roared off to touch the stars
Hearts and eyes strained in anticipation
At this great feat of man
Conquering space.
And suddenly before our very eyes
A deadly flame ignited.
Instantly
Propelling all into eternity
Our hopes dissolved in worms of smoke.
Gone was our joy, our jubilation
Horror and then
Sadness dropped out of the sky
Upon our world - and seas
Into a million pieces.

Alma Foggo York
Huntsville, Alabama

My Thoughts About Space

When my teacher (Mrs. Oaks) said that Challenger had exploded I nearly fainted of shock. I've always loved space. I even tried to build the Challenger with my legos.

I felt bad
and very sad.
The future belongs
to the brave,
and not the afraid.

Jarrod Dodson
Grade 4
Roger Chaffee Elementary School
Huntsville, Alabama

1/29/86

Dear Sir,

The explosion of Space Shuttle Challenger and the reported death of the crew members, rather Heroes, has been received with profound sorrow in Pakistan.

The tragic loss of the Space Shuttle will be felt keenly around the world by all those who have even remotely followed man's attempts to unravel the mysteries of space. The destruction of Challenger and the associated loss of life is a set back to the Shuttle program no doubt, but it is hoped that scientists will be able to cope with it and not permit it to weigh down the entire schedule.

The names of the crew of Challenger's unfortunate 51-L Space Mission, i.e. Francis R. Scobee, Ronald McNair, Michael J. Smith, Gregory B. Jarvis, Ellison S. Onizuka, Judith A. Resnik and Christa McAuliffe, will be remembered forever in the History of Space and Man.

God Bless Them.

Please accept me and my country's heartfelt condolences.

I am closing my letter with President Ronald Reagan's words:

"The Future Does Not Belong To The Fainthearted.
It Belongs To The Brave."

Thanking You,
Yours Sincerely,

Mateen Hamza
(Science Student)
Karachi
Pakistan

1/31/86

Dear Sir,

You cannot believe how upset I was on hearing of the destruction of the space shuttle "Challenger" and her gallant crew. It is truly a tragedy that has occurred.

I wish to offer my sympathies to the administration, and to the families of the crew - indeed, to all Americans who feel as I do.

I believe the shuttle is the way of the future, and hope it will survive for a long while to come.

I hope to write again, in better times.

Yours faithfully,

Jim Leary
Auckland, New Zealand

Power Off

Power on.......
Pushing in the video tape,
Watching my seventh graders,
Waiting for response to Dionne and friends.

Excitement and enthusiasm
Recording the song about friendship,
Planning the Spring poetry unit,
Surprising my students with the video.

Looking upon smiles, bright eyes,
Welcoming Dionne and friends,
Drinking every word,
Feeling each note,
Discovering the heart's openness.

Reading, singing, and digesting the words,
Experiencing poetry through that special closeness.
"Knowing you can always count on me for sure;
That's what friends are for."

A teacher friend slowly walking into the classroom,
Whispering painfully, "I think the Shuttle didn't make it."
Disbelieving, mind wheeling,
Escaping to Dionne and friends.

Video ends.
A dead silence,
Except for TV buzz.
A feeling of helplessness, hopelessness,
I look upon
My little, lost friends,
Big, wetting eyes,
Pounding hearts,
Frowning faces,
Feeling,
Wondering.

The words of our song become real,
Speaking and embedding meaning
That will always live with us -
Dedicating to the astronauts.

Silence.
Starting the video again and knowing that together we can,
"Keep smilin', keep shinin',
For good times and bad times........."

Power off.........

C. Arszman
Middle School English
Academy of Academics and Arts
Huntsville, Alabama

I was at home, sick for the day. It interrupted my TV program. It was an ABC Special Report. I thought, "Darn!" Another one of those news bulletins about something that's going on halfway across the world and doesn't affect me." Well, it did. As those red and black words disappeared from the screen, and Steve Bell's worried, confused and astonished face appeared in their place, I sat up on the couch. The first four words this man uttered out of his pale lips touched a place in my heart that very few things do. My first reaction was confusion. Many thoughts ran through my head, but an interesting one was, "What space shuttle blew up? I didn't even know one was going up!" I am ashamed of that thought, because it means that I take for granted the space shuttle program and the astronauts. But I pushed that thought out of my head, and tried to concentrate on what Steve Bell and Peter Jennings were saying. I had knots in my stomach and tears in my eyes as they talked about and showed pictures of this awful tragedy. As I watched, I thought about my mom at work and that she probably didn't know yet. I thoughtfully slid off the couch and walked to the phone. As I heard mom say "Hello?" I heard myself say in a voice much like Steve Bell's, "Mom, the space shuttle has exploded and......."

Brooke Judkins
Mt. Gap Middle School
Huntsville, Alabama

SIXTH GRADE SPACE CLASS
ACADEMY FOR ACADEMICS & ARTS
2800 POPLAR AVENUE
HUNTSVILLE, ALABAMA 35816

March 3, 1986

His Excellency Mikhel S. Gorbachev
General Secretary of Central Committee
of the Communist Party of Soviet Union
Kremlin in Moscow

Dear Secretary Gorbachev:

We appreciate how you reacted in sorrow instead of in an advantage. We wish you would join in space technology with President Reagan. We have sent a Warm Fuzzy and a poem for this occasion. Thank you and the Soviet Union for sending the letter which could be the key to the door of peace.

Would you please write to President Reagan about what we have said? We are two eleven year olds and one ten year old. We attend a talented and gifted program at a fine arts school, The Academy for Academics and Arts. We were supposed to study this shuttle mission closely because the first teacher in space was on the crew. This tragedy has affected all of the students in our school. We would like for you to communicate this to the children of your country about all of this. We have enclosed pictures of us, the story of Warm Fuzzies, and a copy of the poem for you.

Thoughtfully,

Russ Stanners
Daniel Feld
John Thompson

Enclosures

The Key to Peace

The key to peace begins with friendship
Unfortunately the start of our possible friendship
Had to start with death.

But seven lives helped the locksmith
Named God make the key that can open the door
To the light called peace.

Peace is the light that has power over all.

John Thompson
Grade 6
Academy for Academics & Arts
Huntsville, Alabama

SIXTH GRADE SPACE CLASS
ACADEMY FOR ACADEMICS & ARTS
2800 POPLAR AVENUE
HUNTSVILLE, ALABAMA 35816

March 4, 1986

President Ronald Reagan
The White House
1600 Pennsylvania Avenue
Washington, D.C.

Dear President Reagan:

We are students in a talented and gifted program at the Academy for Academics and Arts in Huntsville. We are writing you to tell you how nice it was for you to make those two wonderful speeches in Washington and Houston. The speeches about the Challenger were very heart-warming and inspiring to everyone.

Here at the Academy we are publishing a book about the Challenger. It will include original art work and creative writing about the Challenger and its crew. We are asking for your permission to publish your two speeches in this booklet. May we please have a copy of each one?

Enclosed is a Warm Fuzzy for you and a copy of a story about it. We wrote a letter to Secretary Gorbachev and that is enclosed as well.

Sincerely,

Candice Black
Cara Britt
Stephanie McGuire

THE WHITE HOUSE

WASHINGTON

April 14, 1986

Dear Students:

Thank you for sharing your thoughts with me. Mrs. Reagan and I were deeply saddened by the loss of the CHALLENGER and its seven brave crew members. We know we share this pain with all Americans and especially with the families who are left to mourn.

I know that the students and teachers of this nation felt a double loss in the presence on this mission of Christa McAuliffe to teach a lesson from space. Although tragedy struck, she has nevertheless taught us all something very important, that the future belongs not to the fainthearted but to the brave. The CHALLENGER crew was pulling us into the future, and we'll continue to follow them with more shuttle launches, more astronauts, more citizen-volunteers, and more exploration.

Perhaps the best way to honor Christa and the other crew members would be to work hard in school and in your community so that someday you will reach your goals and dreams as she wished you would. God bless you.

Sincerely,

Ronald Reagan

Students of the Sixth Grade Space Class
Academy for Academics and Arts
2800 Poplar Avenue
Huntsville, Alabama 35816

Reach For The Stars

The majestic rocket
Reaching toward the sky
Seven aboard
Soon to be none.

The people all watching,
Thunderstruck and pained
Could not believe it
There were none.

72 seconds...
Before the mighty bird
Exploded in her fury
Taking seven people with it.

A nation
was at a standstill
Everything was so tranquil
And the echoing of engines
Was loud and lifelike.

Seven people
To reach the heavens
It couldn't be true
It just couldn't be true.

A twinkling star
In the still night
Seemed to gleam
In memory of the seven heroes.

Mark Farbman
Grade 8
Academy for Academics & Arts
Huntsville, Alabama

1/29/86

Dear Sir,

It's heartening and self-satisfying to write you for the very first time. Greetings and good wishes to you all.

I am a 19 year old Nigeria Science student and an unabashed admirer of the U.S. Space Program especially the manned Space Shuttles - Challenger, Columbus, and Discovery. I like Space Exploration the American style.

I have followed with keen interest and admiration the U.S. Space Program through the Electronic and Print Media here in Lagos and in the U.S., especially through the "Voice of America" broadcasts to Africa. We are shocked and pained to the core to learn of the explosion of the Challenger shortly after take-off, killing those seven precious, fine U.S. Astronauts. We share in your grief. We mourn with you the loss of those seven superb human beings of High Tech. We urge NASA and the entire American people to take heart and bear the loss with fortitude. We have great respect and admiration for the U.S. Space Program and we urge you guys to go ahead. God bless you.

Best wishes and kind regards.

Cordially,

Loveday Christian Nwankwonta
Lagos, Nigeria
West Africa

2/3/86

To Whom It May Concern

My name is Justin Pickrel, I attend Churchill Jr. High in Galesburg, Illinois.

Our yearly science fair is coming up in March and I decided to do a project on the Space Shuttle.

I've heard some kids knocking the shuttle after Challenger exploded, really putting it down. I've also heard some adults do the same.

I would appreciate it very much if you could send me all the information you have on the shuttle and anything connected with the shuttle so I will be able to show, and explain to these kids and adults, how important the shuttle program is to our future.

Thank You Very Much,

Justin Pickrel
Grade 7
Churchill Jr. High
Galesburg, Illinois

1/30/86

Dear Sirs,

It is with deep regret that we here in the United Kingdom saw with horror and disbelief the in-flight explosion of the space shuttle "Challenger" on 28th January, 1986.

I only hope and pray that the seven brave astronauts who sadly perished will become a shining example to the never really appreciated heroes of the American Space Programme. I am sure that once the reasons for this tragedy have been determined new milestones and discoveries will be reached.

Yours faithfully,

Vyv Wilkins
Glen Parva
Leicester, England

Dear Sirs,

In reference to the recent tragedy concerning the Shuttle Craft Challenger, I would like to extend my deepest sympathy. I know this loss cannot be measured, and it will take time to heal, but with the help of God you will continue.

Till this sad event I had become very complacent and uninterested in space exploration. I guess due to the many achievements and accomplishments made in space I have felt that it is an environment not unlike the one we function in here on earth. This perception I know is not correct. Space is still a frontier not unlike the ones our forefathers encountered in the New World. My son and I have since the accident begun research into the Shuttle Program in order that we might become more aware of the significance in our everyday lives.

Once again, I would like to extend my deepest sympathy. Also I would like to suggest that the next shuttle orbiter, if one is constructed to replace the Challenger, be named SPIRIT. This in memory of those who have lost their lives and because The Spirit does live on in all of us even though the ones we love have gone. So continue the good work and May God Bless each and everyone of you.

Sincerely,

Robert Lee Plemmons, Jr.
Nesbit, Mississippi

2/12/86

Dear Sir or Madam,

I am very sorry I am just writing this letter, I just got your address. I give my deepest sympathy to the families and you. I kept thinking about what if that was my mom and dad.

I hope you find out what happened.

Sincerely,

Malinda Hill
Grade 5
Gardendale Elementary School
New Castle, Alabama

Dear Sir,

I am very sorry that your space craft did not make it into space. At school our project is space. We would like to know much more of space if we could. I am very sorry for the children of America because they were probably just as sad as everybody is. We are watching a TV programme called *The Boy From Space.* It is very exciting. We have been making models. The one that we are doing now is a space buggy made out of paper.

Yours Sincerely,

Claire White
Callington Primary School
Callington, Cornwall
England

Dear NASA Center,

I form part of a space group called "The B'Kara Space Group" which meets every week in the school hall. We are 40 in our group all students of about 15, 16 and 17 old of age.

I am very sorry for the tragic accident that happened to Space Shuttle Challenger. I give my condolences to the families of the victims of the accident which as President Reagan said are heroes of mankind because of the part they were giving for mankind in the building of a space station.

Yours Faithfully,

Silvano Azzopardi
Censu Busuttil Str.
B'Kara
Malta
Europe.

Dear Sir,

The original intention of this letter was to ask for some information about recent shuttle flights, but after the terrible tragedy of STS-25 I decided that a letter of sympathy would be far more appropriate.

I have followed the shuttle programme intensely ever since Enterprise was rolled out. I decided that once the whole fleet was in operation, I would choose one of the shuttles as "mine". This would be the craft that I would concentrate on most as well as following the progress of the other three. I have four models and three large posters of this orbiter. I expect you have already guessed that the shuttle I chose was "Challenger" - for two reasons - its advanced technology over the others and its reliability.

The events of the 28th January left me devastated as I saw "my" ship just disappear on television. This was the first time I have ever cried over something shown on TV.

I must admit that my first concern was for the Challenger itself but, after seeing films of Christa and the rest of the crew before boarding, my sympathy went to the families left and all the NASA people involved with the programme.

I am just 17 and I really felt for the school children watching the launch and I still cry when I see it now.

I think two ways of honouring the Challenger seven would be to have some form of plaque or other reminder of the mission ejected from the next shuttle to go up. A long term reminder and honour to the crew would be, if a replacement shuttle is built, to name it "Challenger II". Some may say this would be bad luck but I think it would ensure that this tragic mission and its crew are never forgotten. Everytime the craft launched, the crew of Challenger would be thought of.

I still intend to follow the shuttle programme when it is reinstated - with as much enthusiasm as ever. I hope all goes well and successful in the future, and please don't give in to the critics who say unmanned flights would be better - I have every faith in NASA and the shuttle programme. The shuttle is a beautiful craft which I dearly want to see back in space no matter how long it may take.

Once again, I pass on my deepest sympathies to all involved with the shuttle programme and especially the families of the seven who perished. I would be grateful if you pass on this message to all involved. Thank you.

Yours ever faithful,

Jon D. Beeching
Bridge Near Canterbury, Kent
England

McAuliffe's Dream

There was a teacher who had a dream
To fly on a U.S. NASA machine.
There was a count down
From ten to one.
Watching her going up
Was a lot of fun.
She was so high up in the air,
I wondered what was going on up there.
As she with others flew with courage,
Things became a little bit blurrish.
There was a big boom.
There were parts in the air
Flying everywhere.
She was gone.
But, down that bright path
We know she's not alone.
That teacher found her dream.
The day she flew with the NASA team.

Christopher Langham
Grade 8
Academy for Academics & Arts
Huntsville, Alabama

I think of this not as a failure but as a chance for a new start. It really has affected me to reach for my dreams.

Tim Hill
Grade 4
Academy for Academics & Arts
Huntsville, Alabama

Keep Reaching

They are heroes
To our nation.
Though they died
To our horror and fright.
We need to keep reaching.

They gave their lives
To see the stars.
To travel where space is black.
Their deaths were terrible indeed.
But we still need to keep reaching.

Time needs to keep going,
The missions keep flowing.
Their untimely death was annoying
But we still need to keep reaching.

Mat Hudson
Grade 7
Academy for Academics & Arts
Huntsville, Alabama

It has helped me to know that sometimes to get something you want, you have to take risks. And sometimes you will not get it, but others will know you've tried.

Rod Montgomery
Grade 4
Academy for Academics & Arts
Huntsville, Alabama

My Feelings About The Challenger I

Last weekend I took a trip to Gatlinburg, Tennessee. I was having such a good time I didn't even think about the Challenger I going up into space. I really didn't care. It was Tuesday, January 28, 1986, that I realized how important and dangerous a space trip can be.

I was riding along listening to the radio in the car. Then all of a sudden a newsbreak broke in to tell us the tragic news about the Challenger I. The newsbreak said the Challenger I had just exploded and gone into two different directions forming a B shape.

It was such a terrifying feeling. It was as if something had stung me, and I was feeling the pain. It struck me as if it was my own friend up there that died for me.

I really felt guilty. I completely misjudged the precious gift of life. I guess some of us take it for granted. I surely did.

This explosion was very tragic but also a good lesson for each and everyone of us. It gave us a chance to realize the precious gift of life.

Stephanie Brunell
Grade 7
Mt. Gap Middle School
Huntsville, Alabama

Dear NASA,

I've been thinking about the space shuttle that exploded. And decided to write.

The seven people that went to explore space had to be brave. I feel sorry for them all especially for Christa M., the first teacher to explore space. I've been thinking about going to explore space. I might be scared but it might be fun.

I also feel sorry for Christa's family and all of the other families too. I saw it on T.V. I also saw the funeral on T.V. I heard the space shuttle blew up at 11:39 a.m.

I think I'm very lucky for the things I have.

Your friend,

Cayci Axley
Grade 3
Riceville Elementary School
Athens, Tennessee

2/3/86

Dear Rod,

I just wanted to drop you a note to let you know that I am thinking of you and your staff, especially the "kids" (counselors) who are planning to be astronauts.

The kids and I have felt this tragedy keenly as we have been talking about the shuttle all year in preparation for our trip to Space Camp in May. We have done a lot of talking this week about having the courage to "go for it" in life and being willing to accept the consequences for the things we decide to do. I am enclosing a copy of a poem one of them wrote.

I don't know whether you have heard from any of the other teachers in my group from the workshop, but as a result of my experiences at Space Camp I am taking this loss personally. I'm sure the fact that Challenger was my ship has something to do with this as well as my being a teacher.

Sincerely,
Elizabeth Hayes
Teacher
Glen Iris Elementary School
Birmingham, Alabama

FLIGHT # 25

Dear to us were the ones
on Challenger #25;
They were the ones
to risk their own lives.

It just seemed so perfect
on that bright and clear day;
The crew, with its teacher
was at last on its way.

We quickly forgot
how risky it was
to send a shuttle
into orbit above.

We were rudely awakened
on that cold winter day;
When it became clear
that it had gone astray.

Our hearts filled with pity;
Our minds with disgust;
The fate of the crew
was very unjust.

It was supposed to be
a very joyous day;
but the earth-shaking blast
blew all of that away.

First time I heard,
I thought it was mock.
But when it set in,
I felt nothing but shock.

No anger, no sorrow,
and no time to mope,
No pain, no horror,
not a flicker of hope.

All that was there
was a split-second flash;
The shuttle had been obliterated
into wind-blown trash.

1.5 million dollars
is not a high cost;
but irreplaceable are
the seven lives lost.

It's all over now
and we must go on;
but we'll never forget
the crew that's now gone.

Greg Greene
Grade 8
Glen Iris Elementary School
Birmingham, Alabama

Dear Mr. Buckbee:

Now that the terrible events of last Tuesday are behind us, I feel even more of an urgency to "get the message of SPACE to the youth of our country". To do that my Scottsdale School District would like to establish a scholarship or two for worthy students in our district. Our Superintendent, Dr. Philip Gates, has issued a directive to all principals in the district.

Please advise me the age or grade of your Space Cadets, cost, and availability of time. Thank you.

Sincerely,

Robin Kline
Arizona Teacher-in-Space
Scottsdale, Arizona

2/18/86

Dear NASA Family,

I feel very mad that the space shuttle explorer blew up. I really wanted it to go up. The most exciting thing in space is that the teachers are going, too. I'm glad the President said that everyday people and not just movie stars and important people like that could go to space.
If I could ever go in space I would be very happy but scared.
I know my family would be happy, too.
I know that the teacher will be remembered and will be in our history.

Sincerely,

Margie Klitz
Grade 4
Dafoe Elementary School
Wautoma, Wisconsin

Grief That Is Not Shown

The tears I feel today,
I'll wait to shed tomorrow.
Though I'll not sleep tonight,
Nor find relief from sorrow.
What I feel I cannot say,
I must keep clearminded.
No today I will not say,
Lest I become tearblinded.

Jennifer G. Webster
Grade 8
Riverton Middle School
Madison County, Alabama

Keep Reaching

Why did you die?
You were so triumphant there
When you flew for us.
You died for us
But you will live through us
Too
And you taught us
How to live our lives
To live our dreams
And to always
Keep reaching for the stars.

Kevin Lee
Grade 8
Academy for Academics & Arts
Huntsville, Alabama

Graphic

David Gropper
Grade 4
Academy for Academics & Arts
Huntsville, Alabama

ROCKET

The seven astronauts were very brave.
We wish their lives that we could save.
They were brave, as brave can be.
They changed the lives of you and me.

Emily Bodnar, Grade 2
Academy for Academics & Arts
Huntsville, Alabama

IV

Tributes

Moment of Silence/ Rainbow of Courage

Moment of Silence

Seven different people
All from seven different places
And from different races
All to do the world a favor,
To fly their shuttle
Into the unknown.
A service for their country
By free will risking their lives,
Never knowing what the future will hold.
Never have they known that they
Wouldn't ever see their families again.
These people were heroes to their country.
The tragedy of their death touched everyone
And made them think about the human race
And brought the world together
In a moment of silence to pay tribute
To our nation's heroes.

Tricia Mills
Grade 9
Grissom High School
Huntsville, Alabama

A RAINBOW OF COURAGE

A rainbow has many colors, just like
The 25th shuttle mission.

A RAINBOW OF COURAGE

On their last voyage, or was it their
First? The courageous crew opened
Everyone's eyes.

A RAINBOW OF COURAGE

Their last voyage on earth, but the
First on their heavenly journey.

A RAINBOW OF COURAGE

The Challenger was filled with many a
Rainbow, spiritual rainbows, ethnic
Rainbows, and rainbows of friendship
To name a few.

A RAINBOW OF COURAGE

On this mission was a "teacher,"
A very special lady. She wanted
To show the children of today, the
World of tomorrow. Just one of
Seven in....

A RAINBOW OF COURAGE

We know them all by name
Dick Scobee
Mike Smith
Judy Resnick
El Onizuka
Ron McNair
Greg Jarvis
Christa McAuliffe
We will remember them always.

A RAINBOW OF COURAGE

Rebecca Presley
Huntsville, Alabama

Dear Mr. Buckbee,

Our daughter, Missy (15 years old), has always loved the Shuttle missions, and has felt a special kinship to the Space Program since attending Space Camp in March, 1984 (when, I might add, she received the Right Stuff award at "graduation"). She is eagerly anticipating her return to Huntsville in April for Level II Camp.

Missy dashed off this "Tribute to the Challenger", in tears, Tuesday night (January 28). We just thought you might like to see it. And we thank you for having touched her life in a very special way through Space Camp.

May our good Lord bless you in your continued vocation of working with our young people.

Sincerely,

Kay and Al St. Pierre
New Fairfield, Connecticut

Tribute To The Challenger

On a journey to the stars today,
The 25th of many,
Our great spaceship the Challenger
Met its destiny.
A seven man crew rode aboard the ship
Five men, two women were they,
Those brave, unsuspecting few
Who saw God's face today.
The liftoff was a mighty feat,
Beautiful and straight,
But one minute after this occurred
The courageous met their fate.
What is the purpose of such a thing
Why must this disaster
Occur when all were unaware
Of the fiery blaze thereafter?
Those seven wonderful men and women
Are now on another trip
To the great space beyond our own
Where only God's chosen will sit.

Missy St. Pierre
Grade 9
New Fairfield High School
New Fairfield, Connecticut

Challenger, Farewell

The sun broke through the Florida sky,
Today is the day we'll all ask why.

Everyone gathered to see the site,
As seven astronauts take their flight.

A delay on the pad, they seem not to mind,
A new world up there they hope to find.

The count down came, three, two, one,
They soared so high, they touched the sun.

The lift off was perfect, no problems at all,
And 8 miles out it still stood tall.

Throttle up was the last command,
Then an explosion shook the land.

The curls of smoke hung in the wind,
Our saddened hearts will someday mend.

But memories will linger, we'll never forget,
For Challenger the brave we never met.

Together they sat, together they died,
Together we sit, together we cried.

They lived, they laughed, they cried, they died,
We bid farewell Challenger seven and say goodbye.

Janet Scarcelli
Huntsville, Alabama

So Do They

I had faith in our heroes
So did they
I wanted to see the stars and moons
So did they
I wanted to learn more about space
So did they
I even wanted to go in space
So did they
I wanted to grow to love space
So did they
I still have those dreams
So do they

Stacey Crane
Grade 6
Academy for Academics & Arts
Huntsville, Alabama

Dear Sirs,

The Workman High School Space Shuttle Club is currently gathering items to include in their time capsule which will be dedicated in May in honor of Challenger's fallen crew. We would greatly appreciate a donation by your organization of some material or a certain item that you feel best represents our shuttle program, and would be appropriate to put into the capsule.

Sincerely,

Richard Horrmann
Grade 11
President, WHS Space Shuttle Club
La Puente, California

Challenged Dreams

The Challenger lay resting,
Ready to launch from its pad,
As watched by Krista's children,
Her mom and her dad.

The liftoff seemed flawless, cheers came
From every place, as the Space Shuttle
Challenger blasted off into space.

And just in one second, their dreams
Turned to tears, as the Challenger
Exploded, just as no one feared.

The witnesses stood gazingly staring
At the sky, wondering and pondering
Of leftover life. The thought of the
Impossible soon was cherished, as
The announcement came that the
Crew had all perished.

They died for their braveness, these
Daring young souls. They represented
Our nation, America, as a whole.

We admire those seven who
Reached for the stars, who
Died for their dreams, as
Well as for ours.

Thomas Howard
Grade 9
Virgil Grissom High School
Huntsville, Alabama

Keep Reaching. . .

Keep Reaching for the future.
Keep Reaching for the stars.
Keep Reaching in anyway to let them know who we are.

We are Americans and proud as we can be.
I feel for each and everyone, who gave their lives for me.

Rebecca Jean
Grade 8
Academy for Academics & Arts
Huntsville, Alabama

Challenger

Tragic is our loss,
a bigger one not known.
We mourn for lost faith,
lost people, and their souls.

These people are our heroes,
as they were our hope,
for doors unlocked, worlds explored,
the right to know unknown.

I pray, even though
they're dead, that their souls
will live on, from now
until the end of time.

Nicki Pelletier
Mt. Gap Middle School
Huntsville, Alabama

Stepping Stones In A Dream

The dawning of yesterdays sun,
this pilgrimage of ours begun.
It was a proud and happy time,
the Challenger's astral climb.

This shuttle was to be,
our classroom in the sky.
We waited to see,
our teacher from up high.

Seconds into her skyward plight,
she splintered and severed.
A non-survivable, darksome flight,
as children we stood, revered.

Our smiles iced in horror,
we stood like marking stones.
The chilling thaw our sorrow,
its' stinging embrace our truth.

They dared to dream of the places,
they wished to take us to.
With fondness, we miss their faces,
and all they tried to do.

Our mother's love shaded by a tear,
with the sunrise upon the stream.
They now have crossed without fear,
finding stepping stones, in a dream.

Lee Moore
New Market, Alabama

July 20, 1969 A.D.

Man and the imagination coupled with will
is reason for searching
It makes place for the soul
Man has thought and fathoms deep
Extending his mind as could he extend his hands
Man who of history has awaited time
Wondered of the heavens
And awaited his place
Has now set foot through space
And touched another world.

Larry Eckert
Frankenmuth, Michigan

Challenger Seven

They rode their immortal ship
into the clear blue sky of a January morn.
Propelled by the spirit of adventure
and a million pounding hearts.
Rising gracefully on tongues of flame
To become, in a flash, a blossoming star ...
in the clear blue sky of a January morn
There, suspended between heaven and earth,
past and present,
... frozen forever ...
in the clear blue sky of a January morn.

Harold Eckert
Teacher
Grade 6 Science
Frankenmuth, Michigan

Ode To Challenger

Life goes on without you,
Yet at a slower pace.
You were there at the wrong time,
But you were at the right place.

We'll mourn your deaths
For a week or two;
Then it will all be set aside.
NASA will ponder the mystery;
Reagan will take it in his stride.

If your bodies are ever found,
Then funerals will be held.
Trinkets and memorials
Will surely be for sale.

Such will be the fate
Once you go away,
But there'll always be those
Who will remember you
'Till the very last day.

I hope you'll be aware of this,
Oh, Brave Challenger Crew,
For as long as I exist
On this earth we once shared,
I will remember you.

Donovan Manley
Grade 8
Ed White Middle School
Huntsville, Alabama

The Eighth Seat

Making your way among Golden Adventures
You sought out the trail that would bring you fulfillment
Sought out the trail that would lead to your dreams -
Dreams you made real as they challenged your spirit.
Willow-winding paths chosen ever so carefully
So much left to fate (will it deal with us kindly?)

Star-Chasers all, you have captured our vision -
Took us along as you crept ever higher
To seek out the trail that would lead to your dreams -
Though we stayed behind, all our souls rode with you,
You faced the danger as we watched below,
Oh, how could we know?

Cautiously Onward you trekked ever deeper
Into the recesses of life's Golden Caverns
Ever revealing the farther you travelled
The warmer the glimmer of life's Golden Secrets -
Life's Golden Secrets not many can see,
You shared them with me.

Godspeed, Precious Seven, we long to reach upward
To bring you back home and hold you to our breast.
You've tempered our judgement with terrible wisdom -
Wisdom we'll use as we finish your quest.
Star-Chasers all, as I search the night sky,
I know you'll never die.

Richard Bell
Arlington, Virginia

Space Shuttle

In the year of '86
The space shuttle went
Up for its last trip.
The first teacher in space
Christa McAuliffe was on
Board this shuttle.
All America watched,
And to their surprise
The space shuttle
Was not in the skies.
In its place a
Fireball appeared;
Then America started to cry.

Anissa Morton
Grade 8
Academy for Academics & Arts
Huntsville, Alabama

When our brief span
On earth is done
We hope our name
Will carry on.

These seven names
Will stay with grace
As long as man
Reaches toward space.

Bret Witter
Grade 8
Whitesburg Middle School
Huntsville, Alabama

The Challenger's Dream

The sky stood still,
The stars shown bright
As the wind whispered
Through the stillness of night.

The nation in shock,
Confusion and grief.
Could this really happen?
No one in belief.

The flags wave half-mast
For the seven who died
Just seventy-two seconds
Into their flight.

They died great heroes,
The "Challenger" team.
Yet they all died together,
Completing a dream.

Sharlene Spears
Grade 8
Owens Cross Roads
Alabama

When Will Be The Last Good-bye?

She's dead.
She's not coming back.
Why do they keep on?

Thirteen hours.
Her face exploited every minute.
Still, they keep on.

Go away.
Please stop the news.
How dare they keep on?

Try to forget.
Light humor would be nice.
But they keep on.

Memorial service.
But that isn't enough.
Oh, how they keep on.

Next year.
Suffering all over again.
Why do they keep on?

Kimberly Hammond
College Student
Lincoln, Nebraska

I'm so very sad
About the lives that were lost
The men and women were brave
No matter what the cost.

Space work is important
To our great nation
For our future defense
We need the Space Station.

I hope the families
Hearts heal fast
May their pain disappear
And God's comfort to last.

Amy Irvin
Grade 6
Madison School
Madison, Alabama

Dear Mr. and Mrs. McAuliffe,

I am sorry to hear about your daughter. We love you. I hope your daughter is in heaven.

With love,

* Amber Ehler
Grade 2
Westview School
Champaign, Illinois

* Original in braille - Ed.

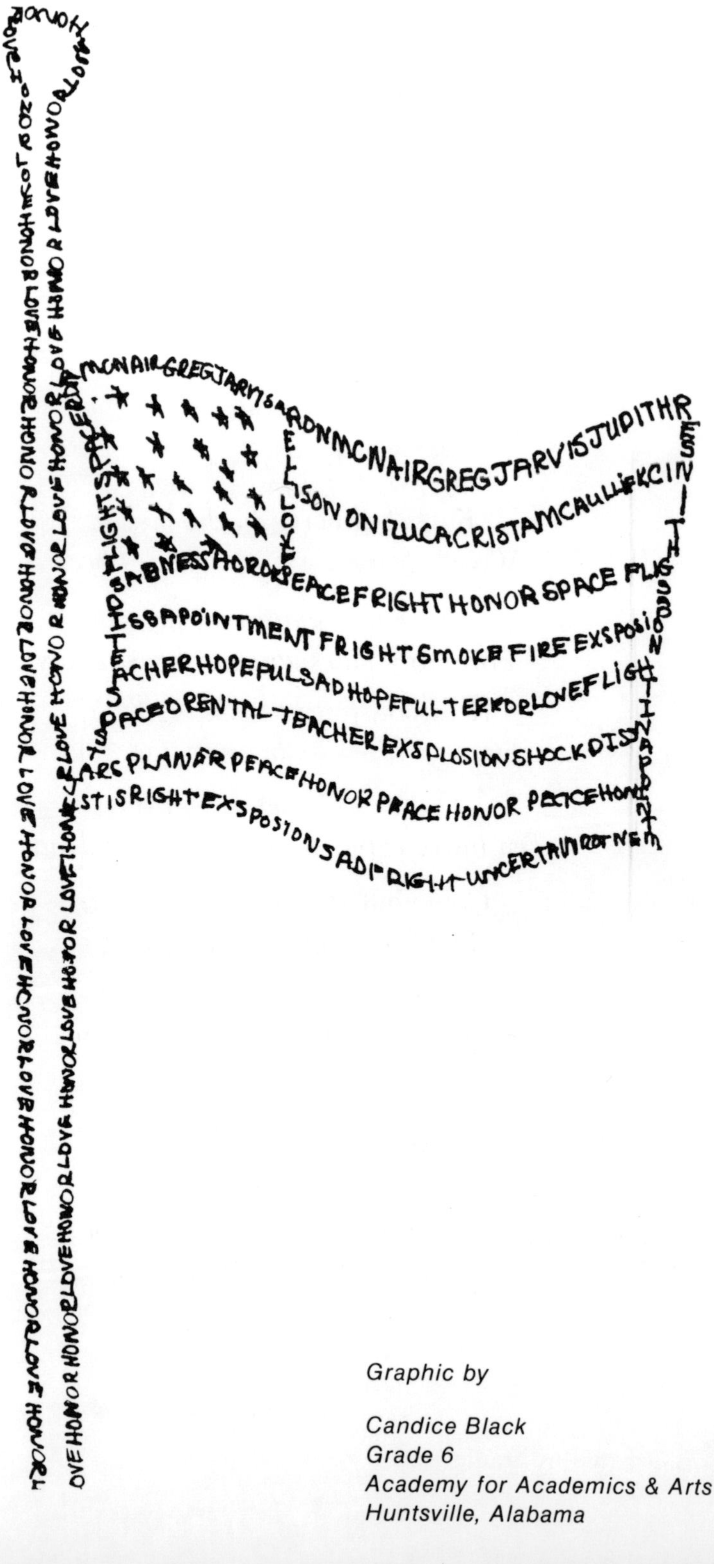

Graphic by

Candice Black
Grade 6
Academy for Academics & Arts
Huntsville, Alabama

My Country

Flags billowing in the breeze
Warm spring days with honey bees.

An eagle soaring in the sky,
For many miles you can hear its cry.

Redwood forests standing bold.
Streams filled with silver and gold.

Halley's Comet over head.
Ten brave explorers who now are dead.

Columbia shooting like a dart.
Freedom rings in every heart.

People coming to be free.
Knowing I have the right to be me.

Ben Laycock
Grade 5
Hunter Elementary School
West Valley City, Utah

Something is missing from the universe
The bravery and courage of seven
There's a gap that will never be filled
Something is missing from the universe
I can picture their smiling faces but
More than memories still remain
Something is missing from the universe
The bravery and courage of seven

Rebecca Thomas
Grade 7
Whitesburg Middle School
Huntsville, Alabama

Billboard on Clinton Overpass
by
Foster and Kleiser Company
Huntsville, Alabama

ACKNOWLEDGMENTS

This book would not have been possible without the support and cooperation of NASA, the Marshall Space Flight Center and the Alabama Space and Rocket Center. Amos Crisp, NASA Public Affairs Officer, and Ed Buckbee and Lee Sentell at the Space and Rocket Center were particularly helpful. Mr. Crisp obtained for our review all the letters written to MSFC. He was instrumental in obtaining NASA's approval for all published letters. Ed Buckbee and Lee Sentell gave us copies of the letters received by the Space Camp and the Space and Rocket Center. Their help has been invaluable.

We appreciate the cooperation of Dr. Mary Jane Caylor and the Huntsville City Schools and Dr. Richard Chapman and the Madison County Schools. Administrators and teachers in both school systems worked with us in acquiring art and poetry from area students.

Thanks to Tom Fricker for providing the drawing of the seven astronauts and to Foster and Kleiser Company for offering the photograph which appears at the end of the book.

A large thank you goes to Rejetta Balentine for the great job she has done in co-ordinating and organizing the massive volume of paperwork which has been generated in the past four months.

We also want to thank Judy and Grady Smith of Great American Graphics, Huntsville, for their guidance and skill in the technical preparation of this book.

Finally, we appreciate the efforts of all those thoughtful young people whose work was submitted but not selected for publication. We would like to have included every single heartfelt letter and poem.

Dottie Main
Jeffrey C. Hindman, M.D.

Publishers Note:

Profits from the sale of *Challenger Remembered* will be used to support the scholarship fund at the United States Space Camp in Huntsville.

JCH
Hometown Press, Inc.